AF575893

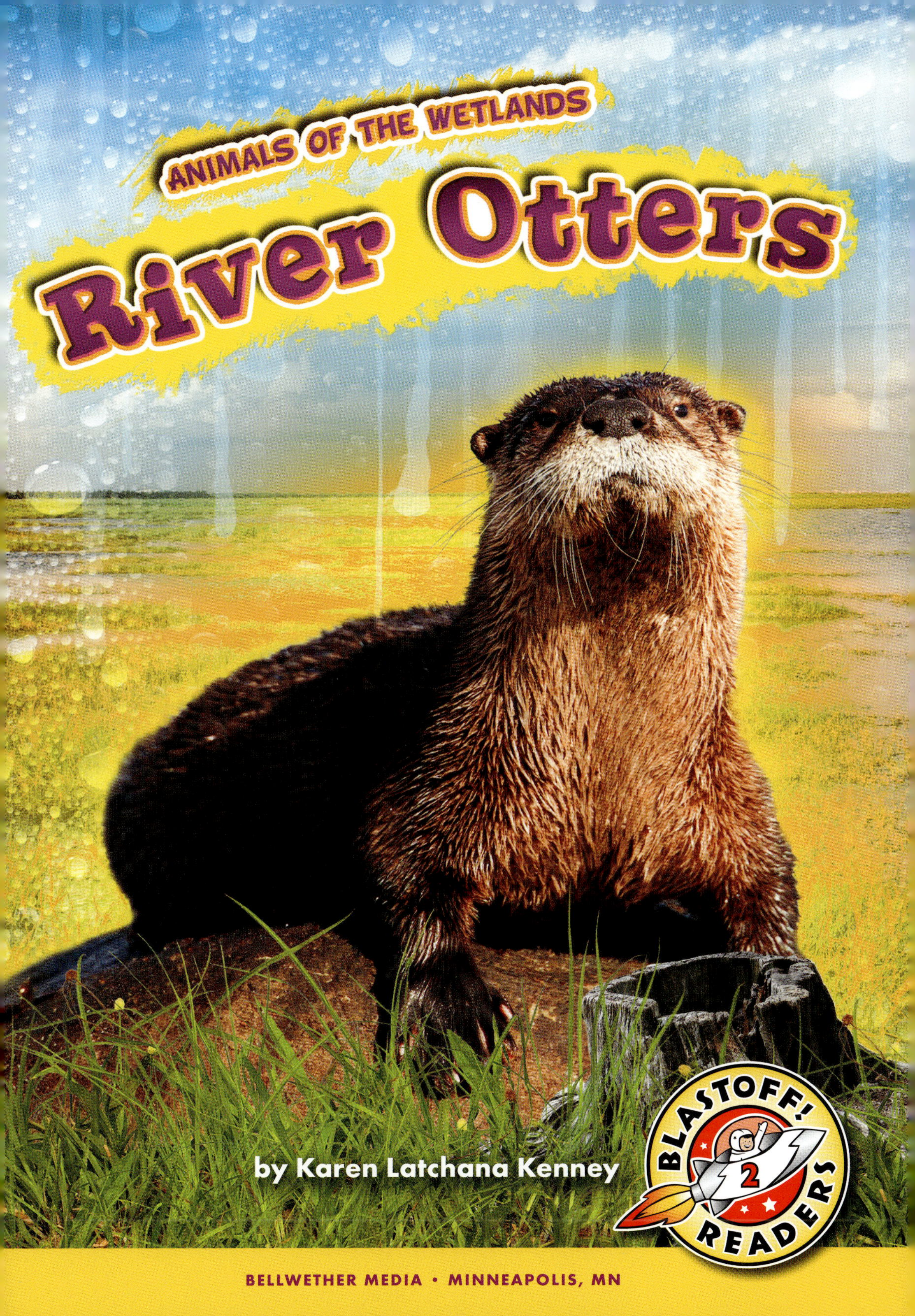
ANIMALS OF THE WETLANDS
River Otters
by Karen Latchana Kenney
BLASTOFF! READERS
2
BELLWETHER MEDIA • MINNEAPOLIS, MN

Blastoff! Readers are carefully developed by literacy experts to build reading stamina and move students toward fluency by combining standards-based content with developmentally appropriate text.

Level 1 provides the most support through repetition of high-frequency words, light text, predictable sentence patterns, and strong visual support.

Level 2 offers early readers a bit more challenge through varied sentences, increased text load, and text-supportive special features.

Level 3 advances early-fluent readers toward fluency through increased text load, less reliance on photos, advancing concepts, longer sentences, and more complex special features.

★ Blastoff! Universe

Reading Level

Grade K

Grades 1–3

Grade 4

This edition first published in 2021 by Bellwether Media, Inc.

Library of Congress Cataloging-in-Publication Data

Names: Kenney, Karen Latchana, author.
Title: River otters / by Karen Latchana Kenney.
Description: Minneapolis, MN : Bellwether Media, Inc., 2021. | Series: Animals of the wetlands | Includes bibliographical references and index. | Audience: Ages 5-8 | Audience: Grades K-1 | Summary: "Relevant images match informative text in this introduction to river otters. Intended for students in kindergarten through third grade"– Provided by publisher.
Identifiers: LCCN 2020033245 (print) | LCCN 2020033246 (ebook) | ISBN 9781644874202 (library binding) | ISBN 9781648340970 (ebook)
Subjects: LCSH: North American river otter–Juvenile literature. | Wetland animals–Juvenile literature.
Classification: LCC QL737.C25 K3965 2021 (print) | LCC QL737.C25 (ebook) | DDC 599.769–dc23
LC record available at https://lccn.loc.gov/2020033245
LC ebook record available at https://lccn.loc.gov/2020033246

Editor: Betsy Rathburn Designer: Josh Brink

Printed in the United States of America, North Mankato, MN.

Table of Contents

Life in the Wetlands

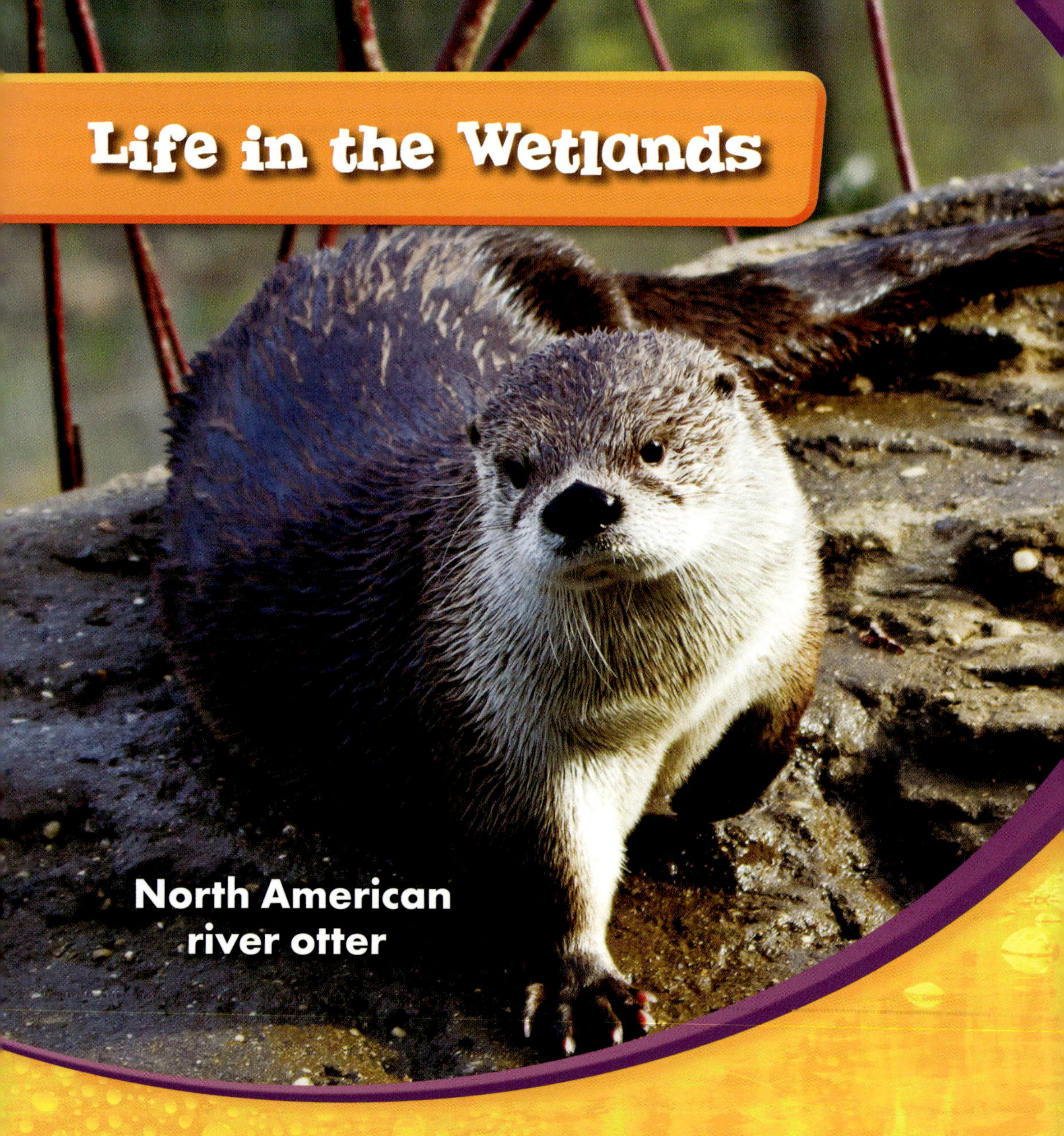
North American river otter

River otters are **mammals** that live in the wetlands **biome**.

They swim through rivers, lakes, and swamps around the world.

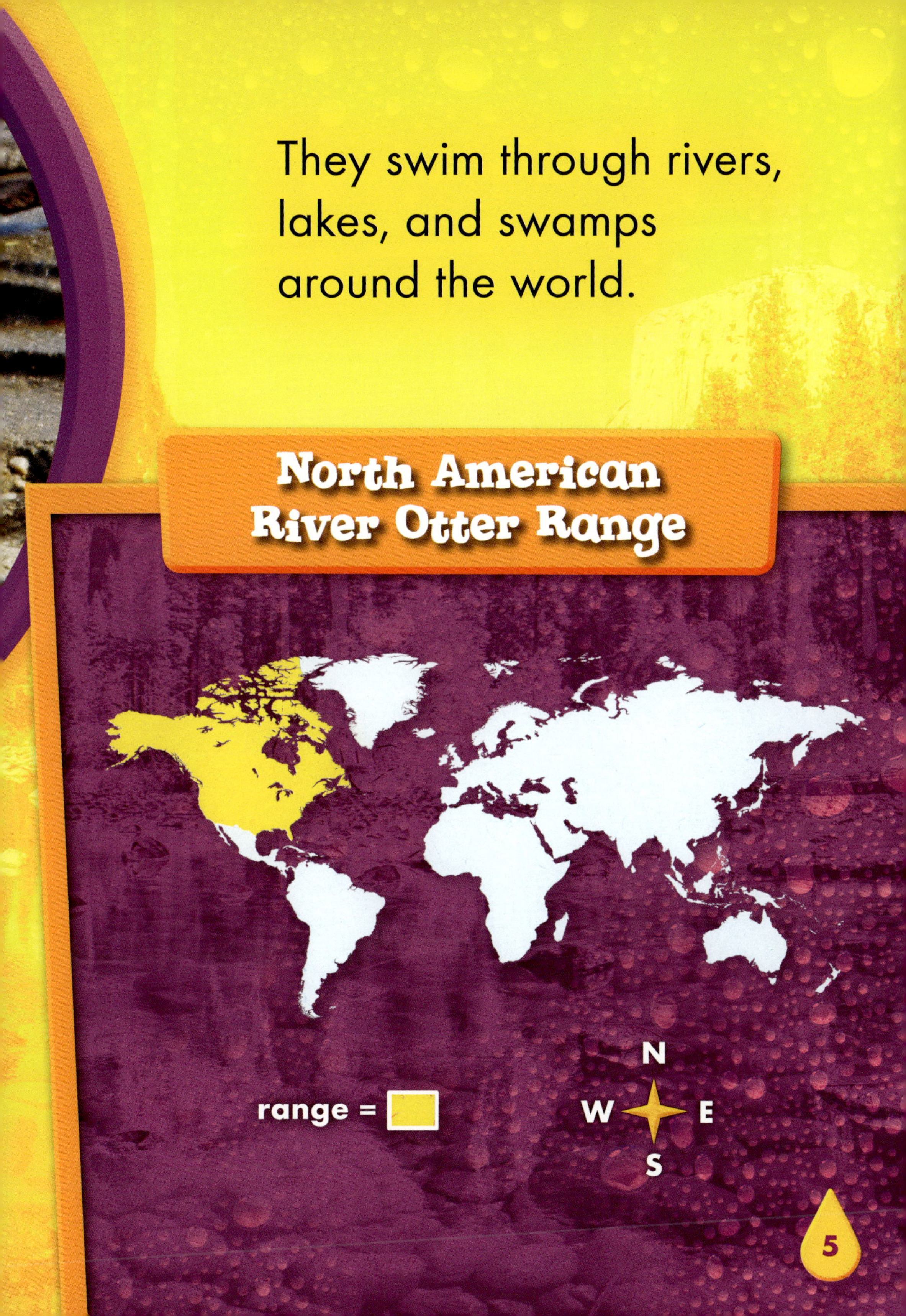

Eurasian river otter

River otters have **adapted** to their wetland homes. Their strong bodies and flat tails help them swim fast.

Webbed feet help them paddle and steer.

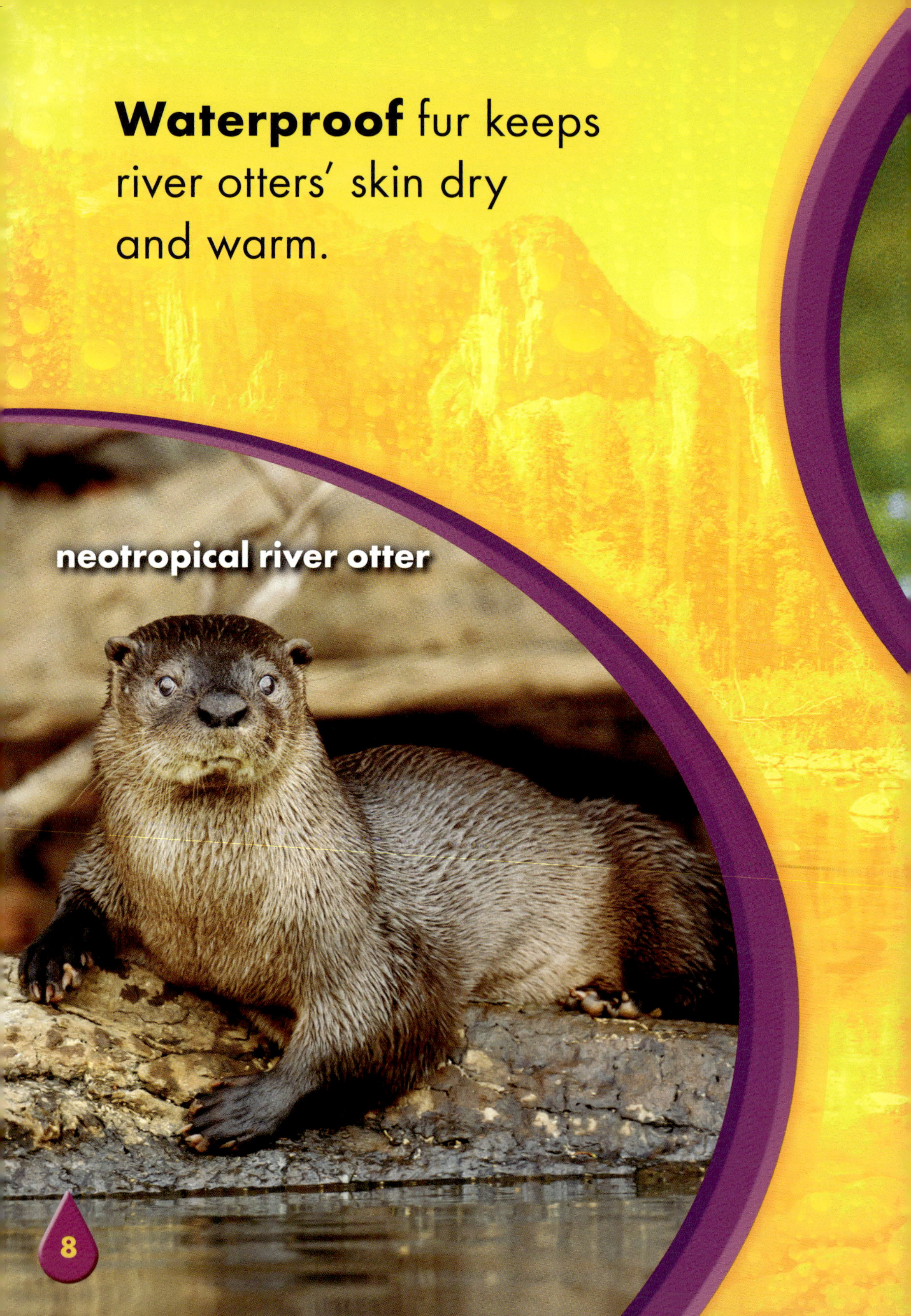

Waterproof fur keeps river otters' skin dry and warm.

neotropical river otter

Long **guard hairs** keep water out. Underneath, short and **dense** hairs trap heat.

These animals must dive for their food. Their ears and **nostrils** close tightly to keep water out.

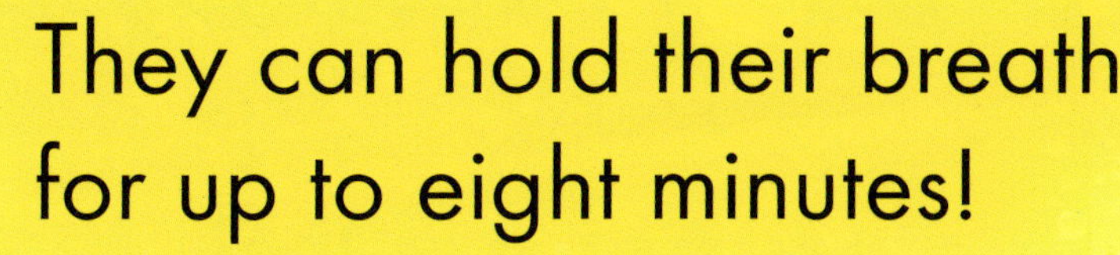

They can hold their breath for up to eight minutes!

Special Adaptations

ears and nostrils that close

waterproof fur

flat tail

webbed feet

Grooming and Playing

River otters **groom** their fur. They rub against grass and sand to remove water.

They spread oil from **glands** under their tails. This keeps their fur waterproof!

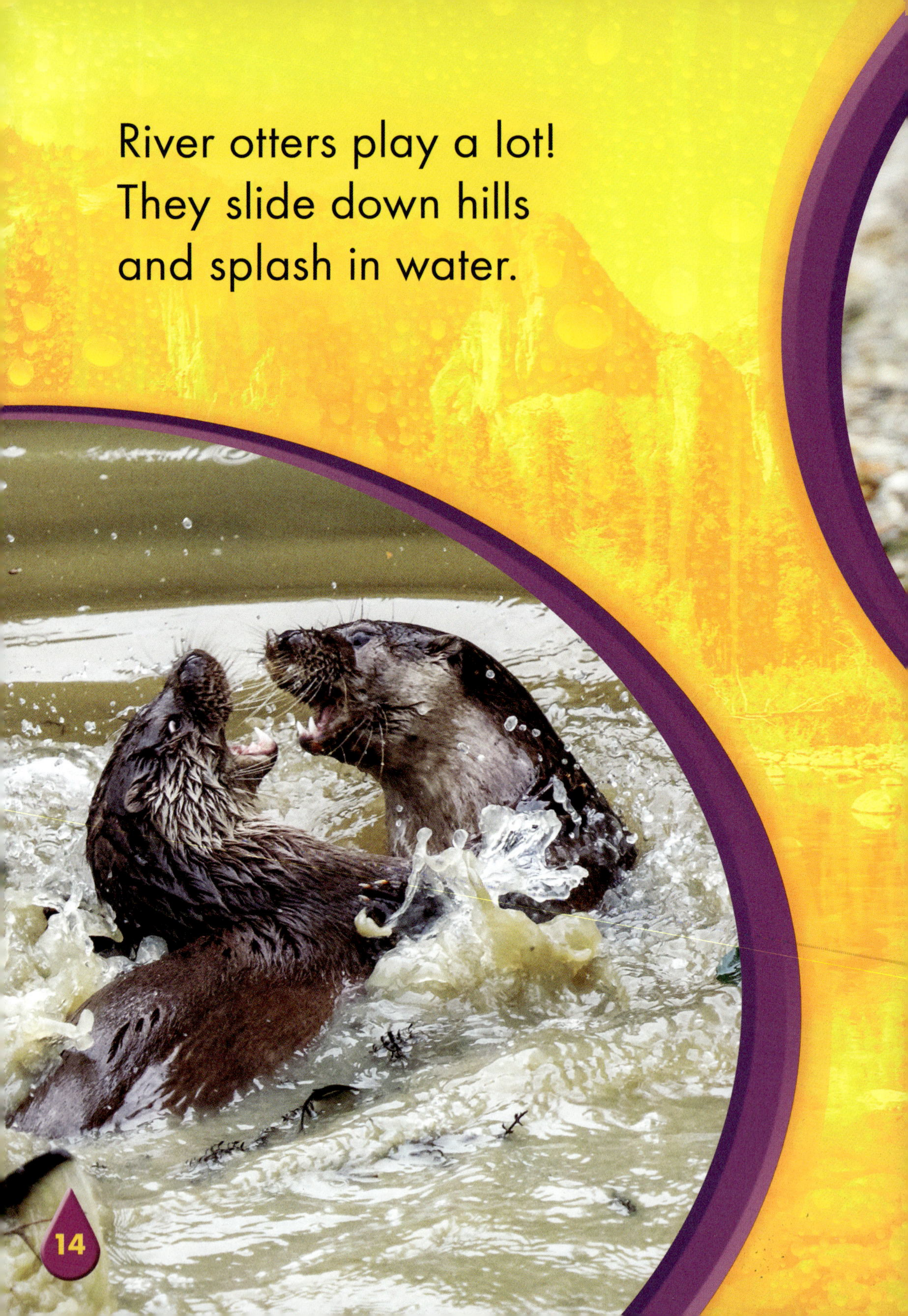

River otters play a lot! They slide down hills and splash in water.

This teaches young otters how to hunt. They learn how to run and chase!

The wetlands can be loud. But river otters have found ways to **communicate**.

They grunt and whistle to each other. They also mark their **territories**.

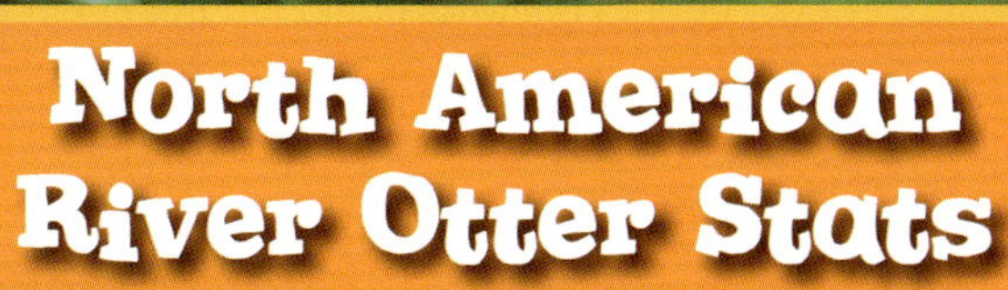

North American River Otter Stats

Least Concern	Near Threatened	Vulnerable	Endangered	Critically Endangered	Extinct in the Wild	Extinct

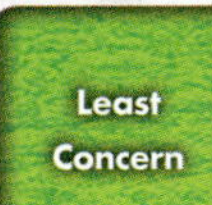

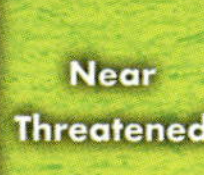

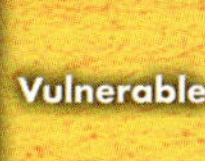

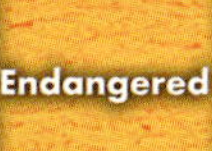

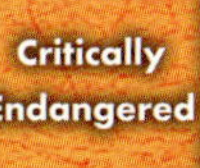

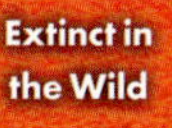

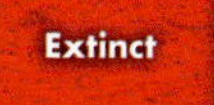

conservation status: least concern

life span: up to 9 years

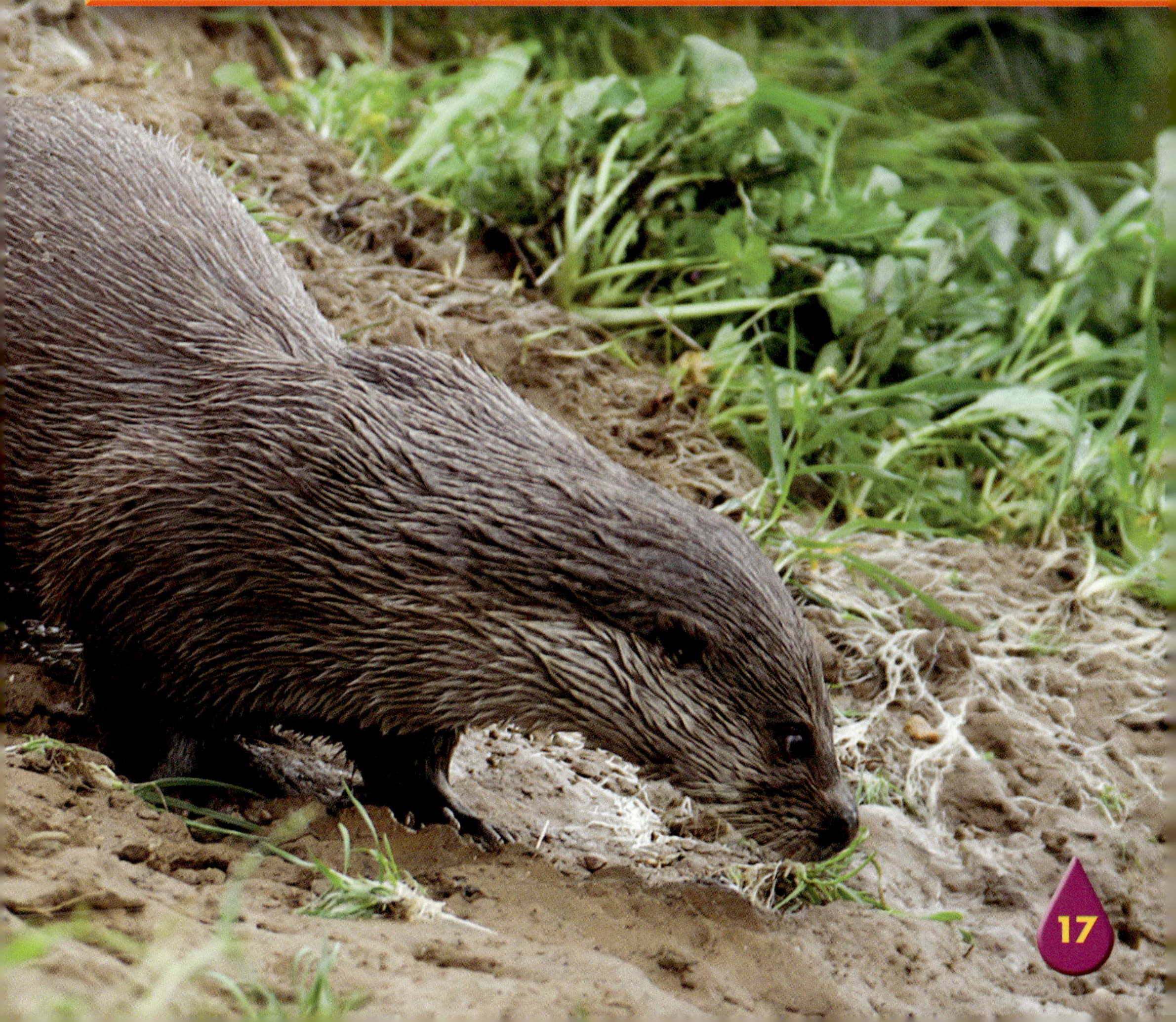

Water Hunters

River otters usually hunt for **prey** at night. Their whiskers sense movement in the water.

Their sharp eyesight helps them spot their next meal!

River Otter Diet

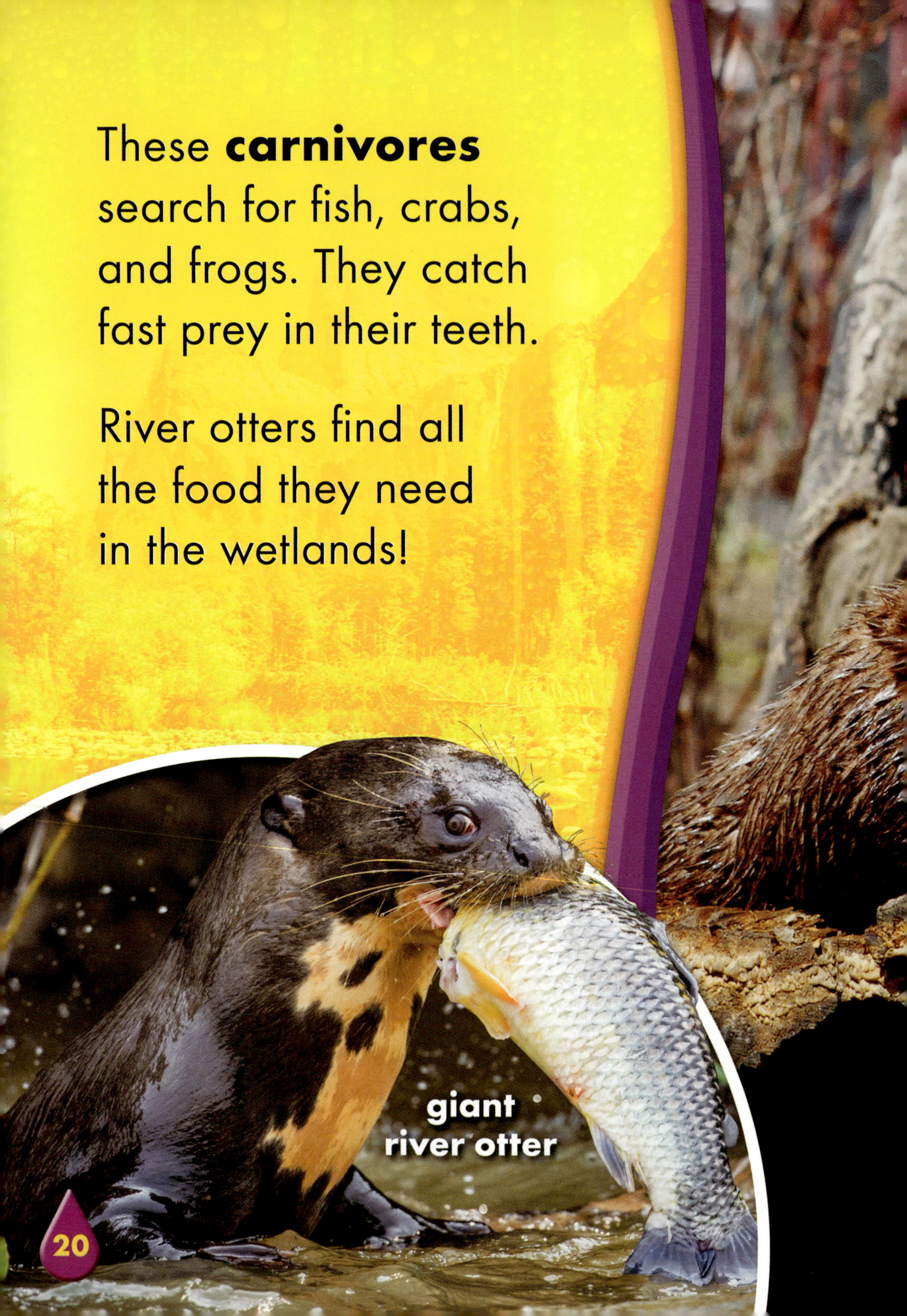

These **carnivores** search for fish, crabs, and frogs. They catch fast prey in their teeth.

River otters find all the food they need in the wetlands!

giant river otter

Glossary

adapted—changed over a long period of time

biome—a large area with certain plants, animals, and weather

carnivores—animals that only eat meat

communicate—to share information and feelings

dense—close together

glands—organs that release oil or other substances

groom—to clean one's fur

guard hairs—long, thick hairs on the outside of an animal's fur

mammals—warm-blooded animals that have backbones and feed their young milk

nostrils—the two openings of the nose

prey—animals that are hunted by other animals for food

territories—land areas where animals live

waterproof—able to resist water

webbed—having an area of skin between the fingers or toes

To Learn More

AT THE LIBRARY

Grack, Rachel. *Beavers.* Minneapolis, Minn.: Bellwether Media, 2019.

Lawrence, Ellen. *North American River Otter.* New York, N.Y.: Bearport Publishing, 2017.

Rake, Jody Sullivan. *Kings of the Rivers.* North Mankato, Minn.: Capstone Press, 2018.

ON THE WEB

FACTSURFER

Factsurfer.com gives you a safe, fun way to find more information.

1. Go to www.factsurfer.com.
2. Enter "river otters" into the search box and click 🔍.
3. Select your book cover to see a list of related content.

Index

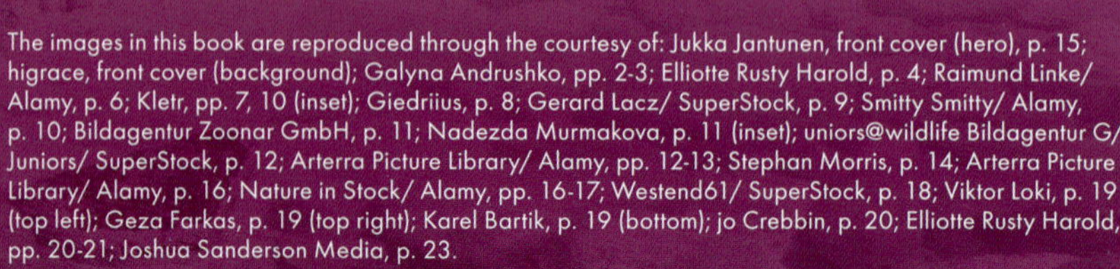

The images in this book are reproduced through the courtesy of: Jukka Jantunen, front cover (hero), p. 15; higrace, front cover (background); Galyna Andrushko, pp. 2-3; Elliotte Rusty Harold, p. 4; Raimund Linke/ Alamy, p. 6; Kletr, pp. 7, 10 (inset); Giedriius, p. 8; Gerard Lacz/ SuperStock, p. 9; Smitty Smitty/ Alamy, p. 10; Bildagentur Zoonar GmbH, p. 11; Nadezda Murmakova, p. 11 (inset); uniors@wildlife Bildagentur G/ Juniors/ SuperStock, p. 12; Arterra Picture Library/ Alamy, pp. 12-13; Stephan Morris, p. 14; Arterra Picture Library/ Alamy, p. 16; Nature in Stock/ Alamy, pp. 16-17; Westend61/ SuperStock, p. 18; Viktor Loki, p. 19 (top left); Geza Farkas, p. 19 (top right); Karel Bartik, p. 19 (bottom); jo Crebbin, p. 20; Elliotte Rusty Harold, pp. 20-21; Joshua Sanderson Media, p. 23.